Earth Day

Jean Feldman and Holly Karapetkova

Tune: *Clap Your Hands Together*

www.rourkeclassroom.com

Earth Day, Earth Day!
Let's all come together.
Earth Day, Earth Day!
To make this world
much better.

'Cause we love our planet Earth,
beautiful and blue.
We want to take care of it
with everything we do.

We can recycle – tell your friends and neighbors! Glass, aluminum, plastic, and paper.

'Cause we love our planet Earth, beautiful and blue.

We want to take care of it with everything we do.

We can plant a tree or two,
to create green spaces.
Walk or ride our bikes
to go different places.

'Cause we love our planet Earth,
beautiful and blue.
We want to take care of it
with everything we do.